Understanding Electromagnetisn

David Drumr

A tutorial for the Electromagnetism component of

OCR and AQA A level Physics

There are of course dozens of text books, revision guides and study aids available. From teaching this subject over 30 years, however, I believe that I communicate my understanding very well and that I am familiar with many of the misconceptions that students have which can hinder their understanding.

I very much hope you enjoy this book and find it useful. Please leave a comment to let me know what you think.

Also currently available as Kindle books or paperbacks on Amazon are:

- Understanding Electricity
- Understanding SHM
- Understanding Mechanics
- Understanding Gravitational and Electric Fields

David Drumm

Before we start

I can help you to tackle exams on electromagnetism, if you're willing to do the work, but for a real understanding of what's going on you need to do some practical work. By building circuits and actually handling components you will internalise the concepts covered here. One of the most useful things is figuring why things don't work. I have learnt a great deal about electricity from pondering over why a particular bulb doesn't light up or why a voltmeter doesn't give the reading that it should.

I am teaching advanced level electricity here, i.e. for A level exams. I will not assume a great deal of prior knowledge however I will assume some mathematical skill, in particular that you can use a calculator. I will also assume that you know that 0.43MJ is the same as 430kJ and the same as 4.3×10^5J. This topic should not be studied without a good understanding of the A level electricity topic.

There will be a lot of quantities with symbols and units you need to learn as well as equations expressing the relationships between them. I will assume that you have the mathematical skill to rearrange these equations as needed. You **MUST** make an effort to learn these equations. You do not have time in an exam to hunt through a formula sheet for an equation that probably isn't in there anyway.

I teach this topic to students over a period of about 3 or 4 weeks so my advice would be to take your time going through this material. Certainly do not go onto a section until you fully understand the one before.

Check your exam board's specification to make sure you are not learning stuff you don't need, e.g. AQA Physics does not include permeance but OCR does.

Contents

What is a B field?	4
Field due to a current	7
Magnetic circuits	8
Force on a current	11
Motors	13
Force on a moving charge	14
Electromagnetic induction	19
Lenz's law	20
Generators	21
RMS values	23
Transformers	25
Questions	28
Answers	32

What is a B field?

A bar magnet produces a magnetic field which we can call a B field.

We can show that there is a magnetic field in a region of space using a small compass called a plotting compass.

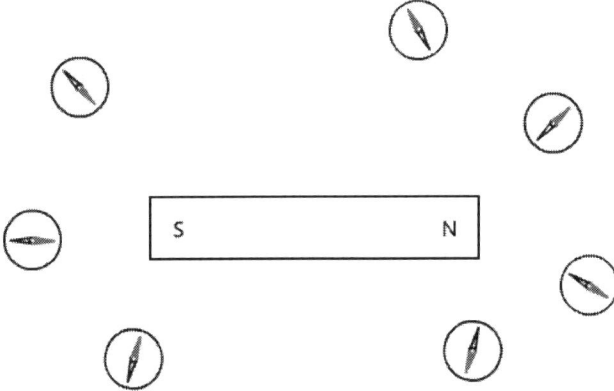

The needle of the compass points towards one end of the magnet and away from the other end. If you sprinkle iron filings around a magnet under a piece of paper you will get a pattern which suggests that there is an invisible field of force surrounding the magnet.

Michael Faraday suggested that it was useful to imagine invisible lines of magnetic force surrounding the magnet. We call these magnetic field lines.

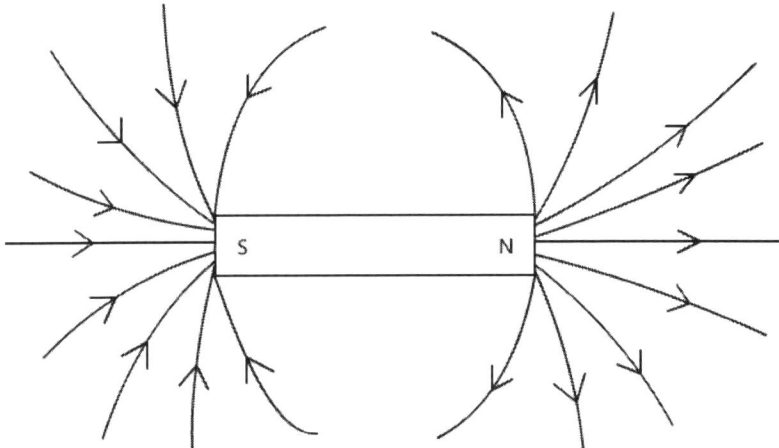

By drawing these lines we not only can represent the direction of the field but how close together the lines are gives us an indication of the strength of the field. The direction is defined as the direction of the force that would act on the north pole of another magnet. As like poles repel each other the field is therefore away from the north pole and towards the south.

Another very useful concept is to imagine that something is flowing out of the north pole and into the south pole. Looking at the diagram above the field appears to be made up of loops of field lines.

We call this quantity magnetic flux which we represent with the symbol ϕ (the Greek letter phi). Its units are Weber (Wb) named after a German scientist called Wilhelm Weber.

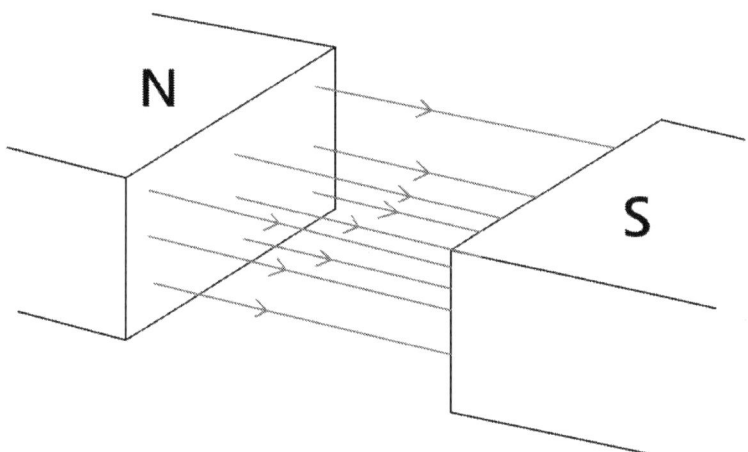

The amount of flux flowing through certain cross sectional area is an indication of the strength of the field.

Magnetic field strength has the symbol B and the strength of a magnetic field is measured in Tesla named after Serbian-American scientist Nikola Tesla.

The field strength B is equal to the amount of flux which flows through a certain area, a quantity we can call flux density.

$$B = \frac{\varphi}{A}$$

Quantity	Symbol	Units	Symbol
Flux	ϕ	Weber	Wb
Magnetic Field Strength or Flux density	B	Tesla or Weber per metre squared	T or Wb/m^2

6

A uniform magnetic field of strength 20mT flows through a coil of diameter 3cm. How much flux flows through the coil?

$\phi = B\,A$

$= 20 \times 10^{-3} \times \pi \,(0.015)^2$

$= 1.41 \times 10^{-5}$ Wb

The vertical component of the Earth's magnetic field at a particular location is 30µT. An airplane with a wingspan of 12m flies through this at 120m/s. How much flux does it cut every second?

$\phi = B\,A$

$= 30 \times 10^{-6} \times (12 \times 120)$

$= 43.2$ mWb

Field due to a current

Hans Christian Oersted discovered that a current carrying conductor produces a magnetic field.

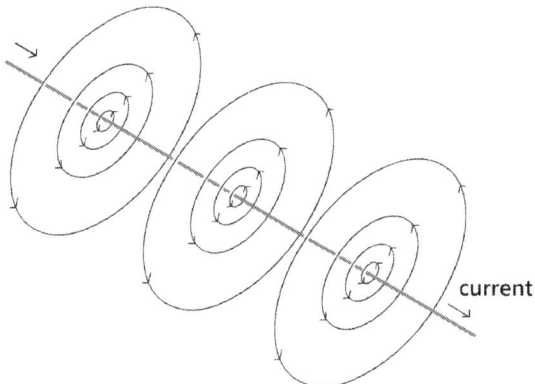

A wire, as in the diagram above, produces a circular field which gets weaker with distance.

The field produced by a wire can be made much stronger if we make it into a coil. Each loop of the coil then contributes to the strength of the field. The field produced by a coil, or solenoid, is very similar to that of a bar magnet outside the coil. Inside the coil, around the middle, the field is uniform.

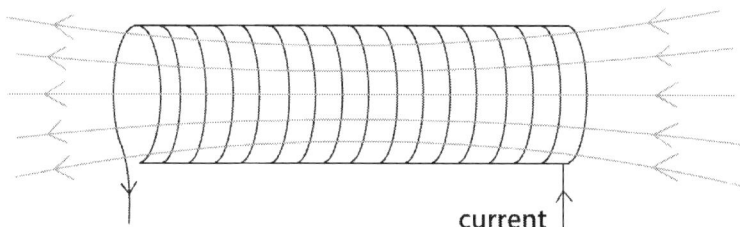

There are a couple of equations which, at time of writing, you don't need to know but may turn up again on a future specification.

A long straight current carrying conductor	A current carrying coil
$B = \mu_0 \dfrac{I}{2\pi r}$ I = current (A) r = distance from the centre of the wire (m)	$B = \mu_0 \dfrac{NI}{L}$ I = current (A) N = number of turns L = length of the coil

The quantity μ_0 is a universal constant called the permeability of free space. We shall learn more about it in the next section. (Free space = a vacuum)

Magnetic circuits

In most electromagnetic machines (motors, generators, transformers etc.) flux flows in a magnetic circuit.

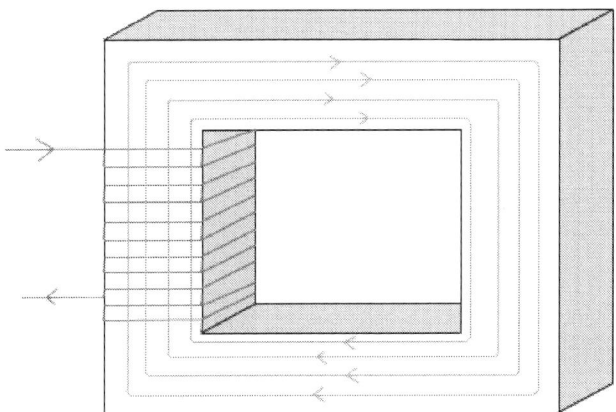

The flux is produced by a current carrying coil. It flows through a core made of iron. Iron is used because it has a high **permeability**. This means that flux flows through it very easily. Because the core is made of iron we will gets lots of flux and the flux will stay in the iron core with very little leakage.

There is a very useful analogy (which you need to know for OCR) between electrical circuits and magnetic circuits.

In both cases something is pushing and something is flowing.

- In an electrical circuit the voltage of the power supply pushes and a current flows around the circuit.
- How much current flows depends on the conductance of the circuit.
- The conductance depends on the dimensions of the wire (length and cross sectional area) and a material property called conductivity.

- In a magnetic circuit current carrying coil pushes and flux flows around the core.
- How much flux flows depends on the permeance of the core.
- The permeance depends on the dimensions of the core (length and cross sectional area) and a material property called permeability.

So permeance is analogous to conductance and permeability is analogous to conductivity. The equations involved are also very similar.

To calculate the conductance of a wire: $G = \dfrac{\sigma A}{L}$

To calculate the permeance of a core: $\Lambda = \dfrac{\mu A}{L}$

To calculate the current in a circuit: $I = G V$

To calculate the flux in a core: $\Phi = \Lambda N I$

The permeability of a material, such as iron, is often expressed in terms of its relative permeability. This is how many times greater it is than μ_0, the permeability of free space.

The value of μ_0 is $4\pi \times 10^{-7}$ H/m

(The unit H is a Henry and is the unit for inductance, not on your spec. at this time)

Air has a relative permeability very close to 1 so for air filled coils we just use μ_0.

Iron, depending on its type, typically has a relative permeability of at least several hundred up to several thousand.

A core is made of iron which has a relative permeability of 500. The core has an internal circumference of 20cm and a cross sectional area of 5cm². A coil of 200 turns is wrapped around the core and a current of 3A is passed through it.

Calculate the following:

a) The permeability of the iron
b) The permeance of the core
c) The flux produced in the core
d) The field strength in the core

$\mu = \mu_r \times \mu_0$ $= 500 \times 4\pi \times 10^{-7}$ $= 6.28 \times 10^{-4}$ H/m

$\Lambda = \dfrac{\mu A}{L}$ $= \dfrac{6.26 \times 10^{-4} \times 5 \times 10^{-4}}{0.2}$ $= 1.57 \times 10^{-6}$ Wb/A turn

$\Phi = \Lambda N I$ $= 1.57 \times 10^{-6} \times 500 \times 3 = 2.36 \times 10^{-3}$ Wb

$B = \dfrac{\Phi}{A} = \dfrac{2.36 \times 10^{-3}}{5 \times 10^{-4}} = 4.71$ T

Sometimes there is a gap in an iron core. The effect of an air gap would be to reduce the permeance of the core as it would be harder for flux to flow around the circuit and there would be much more flux leakage.

There is a quantity which you almost certainly don't need to know, called reluctance. It is the opposite of permeance so would be analogous to resistance. Having an air gap in the circuit would be like putting a large resistor in series in an electrical circuit.

Here's a neat trick

Permeability is the property of a medium that tells us how well a magnetic field flows through it.

The permeability of a vacuum is μ_0, is $4\pi \times 10^{-7}$ H/m

Permittivity is the property of a medium that tells us how well an electric field passes through it.

The permittivity of a vacuum, ε_0, is 8.854×10^{-12} F/m

Multiply these two universal constants together, find the square root of the answer and then find the reciprocal of this.

What other universal constant does this produce?

What has this got to do with electric and magnetic fields passing through vacuums?

James Clark Maxwell was an incredible British scientist who, amongst a multitude of other things, predicted the speed of light in a vacuum well before it was ever measured accurately. If you study physics beyond A level you will come across Maxwell's equations sooner or later.

It was Maxwell who first used the symbol B to represent magnetic field strength.

Force on a current

A current carrying conductor in a magnetic field at right angles to the current experiences a force.

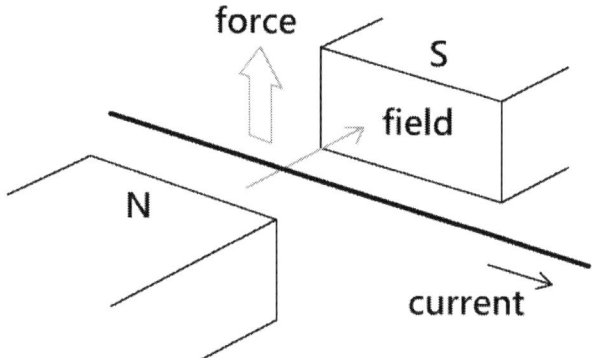

This happens because of the way the field produced by the wire and the other field interact. We can imagine magnetic field lines as rubber bands which have a tendency to straighten. Looking at the diagram below you should see that as there are more field lines on one side of the wire a resultant force will be produced. This is often referred to as the catapult effect.

The ⊕ symbol represents a current flowing into the paper away from us. For a current flowing towards us we would use ⊙.

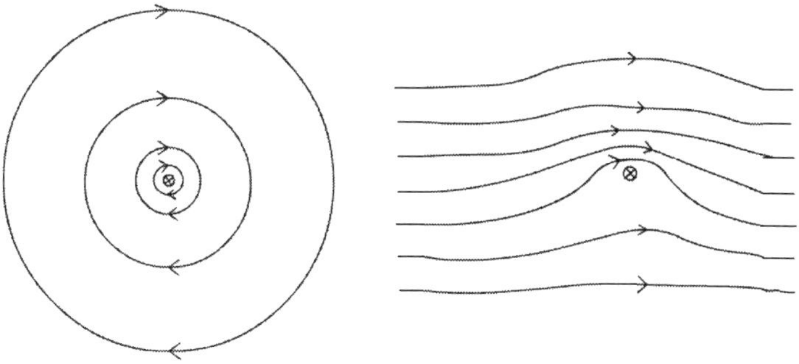

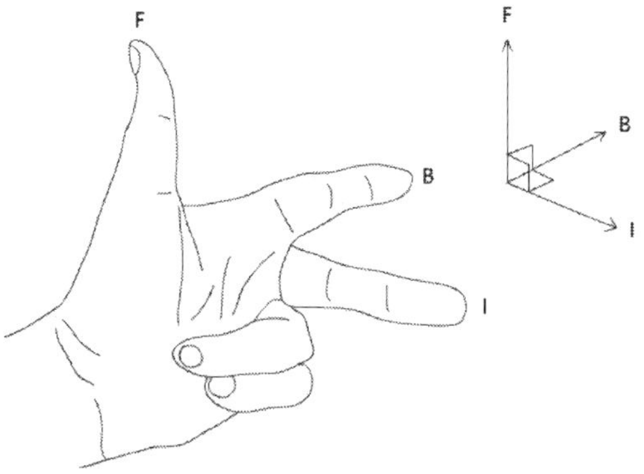

The direction of the force can be found using Fleming's left hand rule.

I suggest that the easiest way to remember this is "starting with the thumb, FBI".

It is easy to show experimentally that the size of the force is proportional to the field strength, the current and the length of the conductor in the field. This may be a required practical on your course.

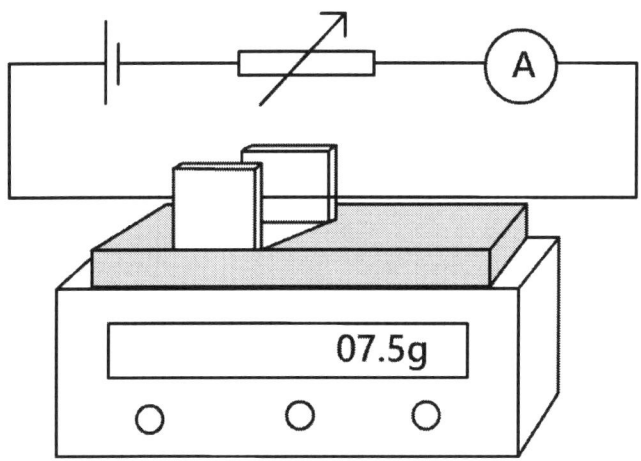

As the wire is pushed upwards then, from Newton's 3rd law, the magnets on the top pan balance will be pushed downwards. From the mass that the balance registers we can calculate the force.

One Tesla is defined as the strength of field required to produce a force of 1 N/m on a wire carrying a current of 1A.

It follows then that: $B = \dfrac{F}{IL}$ and $F = BIL$

If the current and the field are not perpendicular then you would use the component of the current perpendicular to the field, i.e. $I \sin \theta$ where θ is the angle between the current and the field.

A 20cm length of wire is perpendicular to a magnetic field of strength 100mT. Calculate the force which would act on the wire if it carried a current of 3A.

$F = BIL = 100 \times 10^{-3} \times 3 \times 0.2 = 0.06N$

Note that we now have 3 sets of alternative units for magnetic field strength

T, Wb/m² and N/Am

Motors

At time of writing motors aren't on either the OCR or AQA specifications. I'll include a bit about them; however, as I feel they are the most important application of forces on currents.

There are dozens of different designs of motor though they all have something in common:

You put electricity in and it produces movement

More often than not a current flows in a coil in a magnetic field and this produces a turning force on the coil.

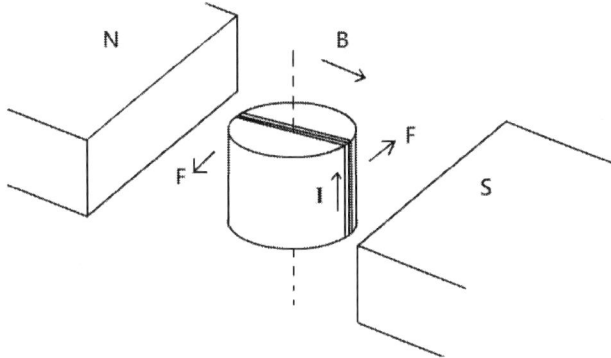

The magnetic field may be produced by permanent magnets or by electromagnets.

Every half turn there is a need to reverse the direction of the current so that the **sense** of the turning force stays the same, i.e. clockwise or anticlockwise. This can be achieved using a device called a commutator or by using an a.c. supply.

There is also a need for some kind of sliding connection between the coil and its power supply so that it can rotate freely. This is another job of the commutator or, in the case of an a.c. motor, slip rings.

The couple, in Nm, acting on the coil is given by:

$$\text{Couple} = B\,I\,A\,N$$

In this equation N is the number of turns of the coil and A is its area.

This equation is just a combination of $F = B\,I\,L$ and $\text{Couple} = F \times d$

If the plane of the coil is at an angle θ to the magnetic field then the equation becomes:

$$\text{Couple} = B\,I\,A\,N\,\sin\theta$$

Force on a moving charge

There are many situations where forces act on charges moving through magnetic fields.

- In cathode ray tubes
- In particle accelerators
- In mass spectrometers

Google a few pictures of the surface of the Sun and you will see charged particles moving in circles due to magnetic forces.

Current is defined as the flow of charge. A moving charge, therefore, constitutes a current and one would expect it to experience a force if it were moving perpendicular to a magnetic field.

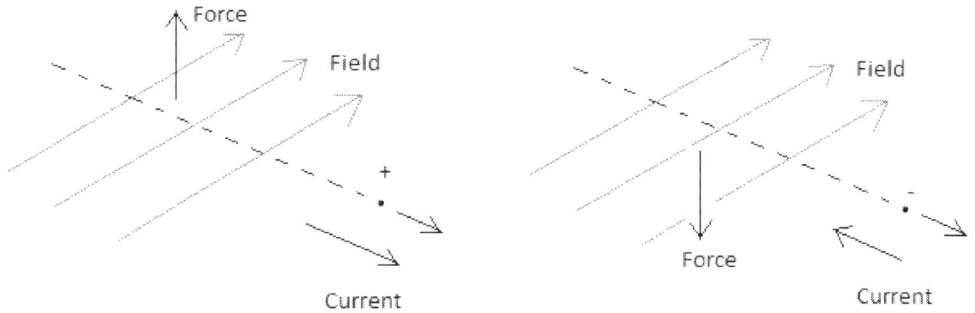

The direction of a current is defined as the direction of flow of positive charge. A positive particle, such as a proton, would experience a force as shown above. For a negative particle, such as an electron, the force would be in the opposite direction.

The size of the force is given by:

F = Bqv

where q is the charge and v is the velocity of the particle.

(If a charge q moves in a straight line at velocity $V = \dfrac{L}{t}$ then substituting this in the equation above gives F = BIL)

Notes:

Unlike in an electric field, a force will only act on the particle if it is moving

Because the force is perpendicular to the velocity the path of the particle will be **circular**, this force being the centripetal force required to move in a circle. The path of a charged particle moving through an electric field is **parabolic**, like a projectile.

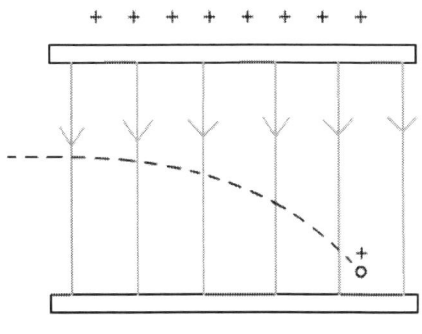 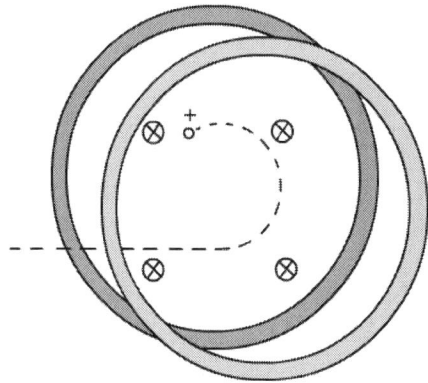

The particle does not gain kinetic energy in a B field. The B field just changes the direction of motion. Because the force and the motion are perpendicular no **work** is done by the field.

A very regular exam question is to show that the radius of the motion is proportional to the momentum of the particle.

$F = ma$ from Newton's 2nd law

$Bqv = m\dfrac{v^2}{r}$

Cancelling and rearranging gives $Bqr = mv$

As B and q are constant $r \propto mv$

Particle accelerators

There are two types of particle accelerator you should know about; the **synchrotron** and the **cyclotron**. There are also linear accelerators but they are not relevant here.

In both types of accelerator electric fields are used to give the particle energy each time it completes a cycle and magnetic fields are used to keep it moving in a circle.

An a.c. voltage is needed to produce the electric field so that the charged particle receives a kick at the right time.

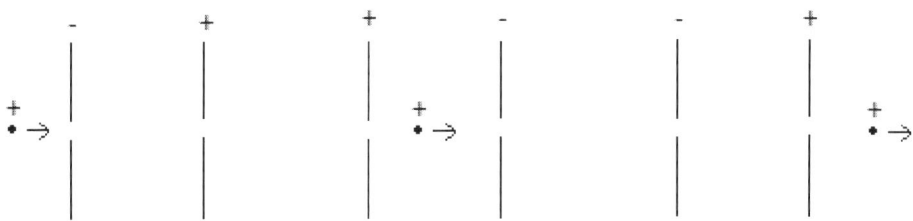

In a synchrotron the radius of the circle stays the same. The electric fields give the particle a "kick" every cycle and so it gets faster and faster.

The electric field needs to be synchronised with the particles to do this, i.e. its frequency needs to increase. The LHC at CERN is an example of a synchrotron.

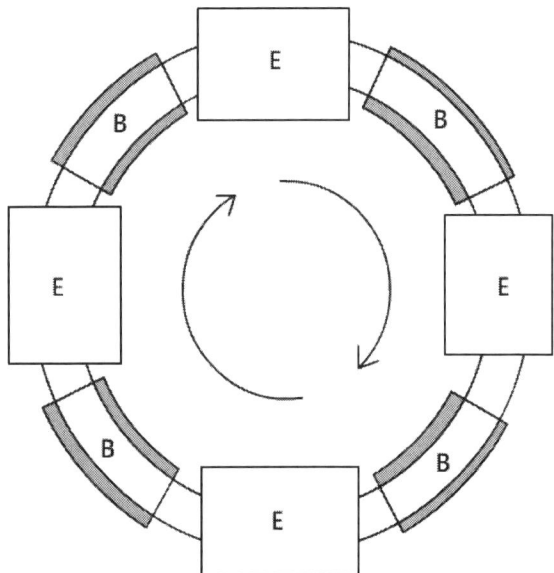

In a cyclotron the frequency of the supply producing the electric field stays the same and so does the angular velocity of the particle. As the particle gets faster the radius of the circle increases.

Smaller accelerators, e.g. in hospitals, tend to be cyclotrons.

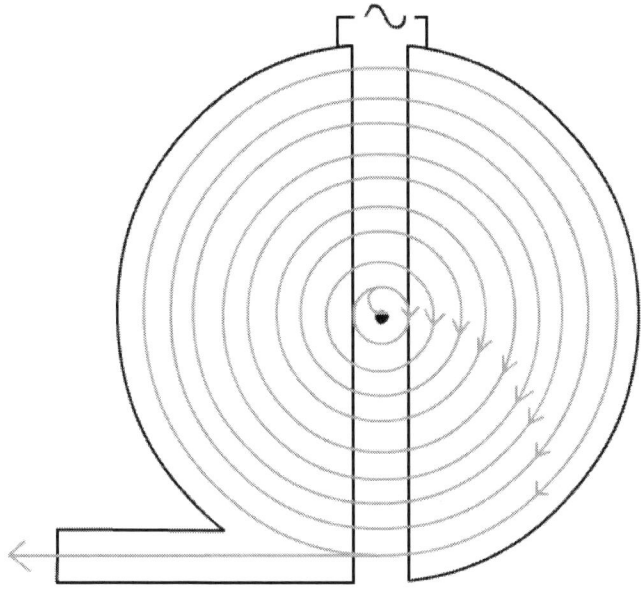

A proton, of mass 1.66×10^{-27} kg, with 200eV of kinetic energy moves at right angles to a uniform field of strength 50mT.

Calculate the following:

- The velocity of the proton
- The force on the proton due to the field
- The radius of the circle it will move in.

$200\text{eV} = 200 \times 1.6 \times 10^{-19} = 3.2 \times 10^{-17} \text{J}$

$v = \sqrt{\dfrac{2\text{K.E.}}{m}} = 196 \times 10^3 \text{ m/s}$

$F = Bqv = 50 \times 10^{-3} \times 1.6 \times 10^{-19} \times 196 \times 10^3 = 1.57 \times 10^{-15} \text{N}$

$r = \dfrac{mv}{Bq} = 40.7\text{mm}$

In particle accelerators such as the ones at CERN the fast moving particles are made to collide with other particles or with each other. They have so much energy that some of this is converted into new particles. Scientists then study the tracks of these new particles in magnetic fields and from the radii of the tracks can deduce much about the charge and mass of the new particles.

The tracks below would be due to the production of a pair of particles with opposite charges and equal masses. As the particles slow down their momentum gets smaller and so the radius of their paths decreases.

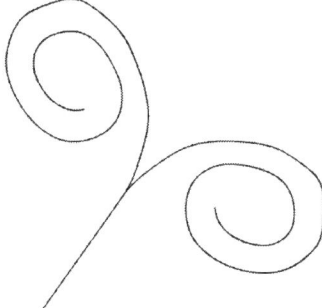

One last application I would like to mention is in one design of fusion reactors called a tokamak.

Incredibly hot plasma is accelerated and contained in a magnetic bottle. When a high enough temperature is reached then nuclear fusion will occur.

Electromagnetic induction

Michael Faraday discovered that you can produce electricity by "cutting" flux with a conductor.

When a conductor is made to move through a magnetic field then an e.m.f. is induced across the ends of the conductor.

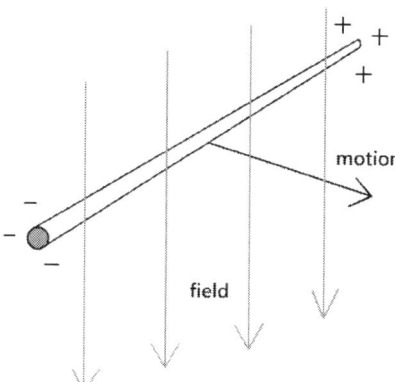

As the conductor moves through the field the mobile charge carriers in the conductor experience a force and move to one end of the conductor. This results in a potential difference across its ends.

Faraday's law tells us that the magnitude of the e.m.f. is proportional to the amount of flux cut per second.

$$\varepsilon \propto \frac{d\Phi}{dt}$$

In S.I. units this equation becomes

$$\varepsilon = \frac{d\Phi}{dt}$$

A straight conductor of length 20cm moves perpendicular to a magnetic field of strength 50mT at a velocity of 5m/s. Calculate the e.m.f. induced across the ends of the conductor.

As $\Phi = B A$ the flux cut per second will equal the area swept out per second x the field strength

$$\varepsilon = \frac{d\Phi}{dt} = B \frac{dA}{dt} = B L v = 50 \times 10^{-3} \times 0.2 \times 5 = 50mV$$

In a generator a coil rotates in a magnetic field. The amount of flux flowing through the coil varies sinusoidally as it rotates and so a sinusoidal e.m.f. is induced.

Lenz's Law

Work must be done to cut flux. If we turn the coils of a dynamo to produce electricity then the work we do is transferred into electrical energy. The induced e.m.f. will therefore oppose us to enable us to do the work.

Here is an analogy.

When you strike a match you scrape the head of the match against the rough surface on the box. A friction force acts against you and so forces you to do work. The work you do is transferred into heat which ignites the match. The induced e.m.f. is like the friction. By acting against us we do work.

Imagine you pushed a strong bar magnet into a coil connected to a bulb. As you pushed the magnet in an e.m.f. would be induced in the coil. As there is a complete circuit a current would be induced which would make the bulb flash.

The e.m.f. would be in a direction that would try to resist you pushing the magnet in, i.e. it would make the end of the coil near the magnet a north pole so that you would have to do work to push the magnet in. If you then pulled the magnet out the bulb would flash again and that end of the coil would become a south pole to resist you pulling the magnet out.

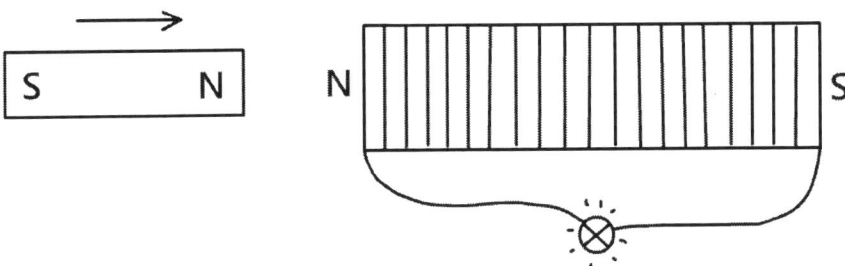

Lenz's law states that the induced e.m.f. will always oppose, or tend to oppose, the flux change causing it.

If there was no circuit then there would still be an e.m.f. but there would be no current induced. Hence the "tend to oppose".

Lenz's law is actually just a form of the principle of conservation of energy.

We modify Faraday's law by adding a minus sign to tell us that the e.m.f. opposes the flux change.

$$\varepsilon = -\frac{d\Phi}{dt}$$

Generators

A generator consists of a coil rotating in a magnetic field (although some designs involve a permanent magnet rotating inside coils).

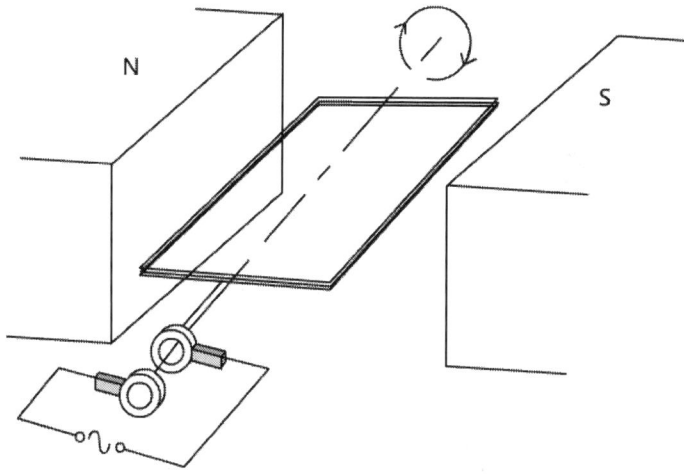

Slip rings and carbon brushes provide a sliding contact to the coil so that it can rotate freely.

The flux linking the coil (the flux linkage) varies with time as follows:

$$N\Phi = N\Phi_0 \sin \omega t$$

Now the e.m.f. is given by

$\varepsilon = -\dfrac{d\Phi}{dt}$ so $\varepsilon = -N\Phi\omega \cos \omega t$

As the maximum value of $\cos \omega t = 1$ we get $\varepsilon_{max} = N\Phi\omega$

A coil with 200 turns and dimensions 7cm by 4cm rotates at 500 r.p.m. in a field of strength 200mT.

Calculate the following:

- The angular velocity of the coil
- The maximum flux linkage
- The maximum e.m.f. induced in the coil

500 r.p.m. $= \dfrac{500 \times 2\pi}{60} = 52.4$ rad/s

$N\phi = N\,B\,A = 200 \times 0.2 \times (0.07 \times 0.04) = 0.112$ Wb turns

$\varepsilon_{max} = N\Phi\omega = 0.112 \times 52.4 = 5.87$ V

The flux linkage and the e.m.f. will vary with time like this, a graph you should learn.

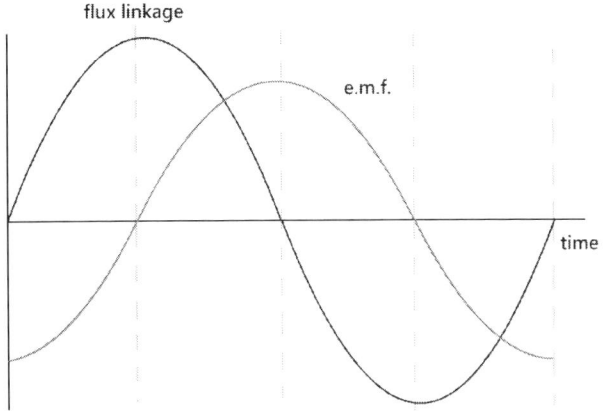

RMS Values

Imagine we had an a.c. power supply which had a peak potential difference of 6V

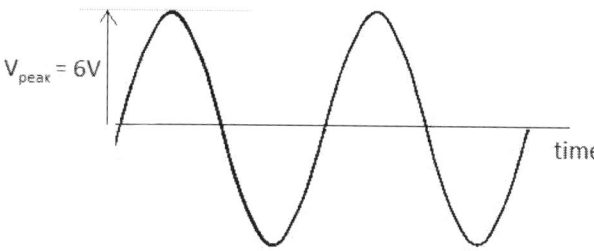

Imagine we connect this power supply across a bulb and another power supply, a 6V d.c. power supply, across an identical bulb.

Would the bulbs be the same brightness?

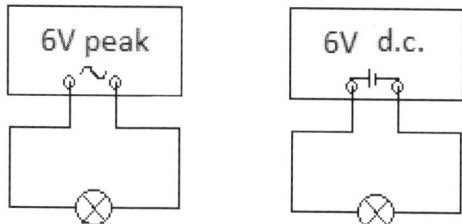

No. The bulb connected across the d.c. supply would be brighter.

To understand why think about what the average voltage across the bulb connected to the a.c. supply would be. It would actually be zero as half the time the voltage is positive and half the time it is negative.

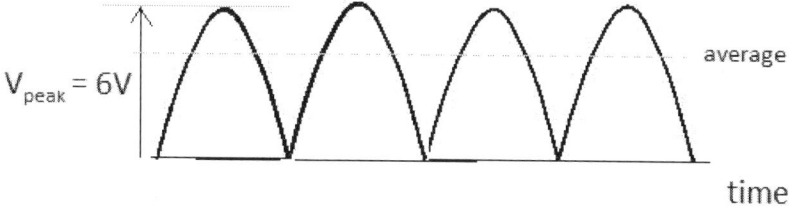

If we "flipped" the negative bits and made them positive then the average would still be less than 6V. For sinusoidal a.c. it would actually be less by a factor of $\sqrt{2}$ which I will prove now.

Consider the case where the power of the two circuits is the same so that the bulbs have the same brightness.

Average power in a.c. circuit = power in d.c. circuit

The voltage at any time in the a.c. circuit is given by

$V = V_{peak} \sin \omega t$ so the power will equal $\dfrac{V^2}{R} = \dfrac{V_{peak}^2}{R} \sin^2 \omega t$

The average value of $\sin^2 \omega t$ is equal to ½ a shown below.

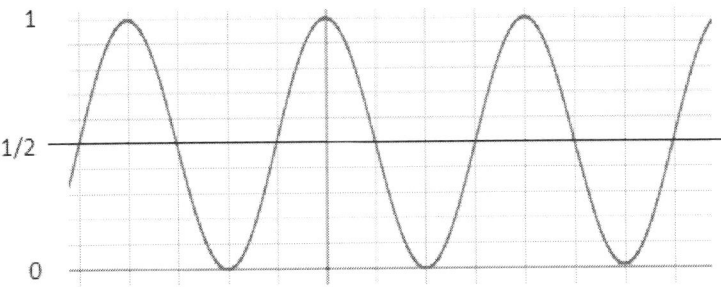

Substituting this in, cancelling the R and taking the square root of both sides gives:

$V = \sqrt{\dfrac{V_{peak}^2}{2}} = \dfrac{V_{peak}}{\sqrt{2}}$

What I have just derived is called the equivalent d.c. value. Another name for it is the r.m.s. value which stands for root mean square. This comes from the fact that we squared the voltage (to get rid of negatives), found the mean then square rooted it.

The a.c. supply in the U.K. is 240V a.c. This is the r.m.s. value (nearly always given when referring to an a.c. supply). The peak voltage from the mains is actually $240 \times \sqrt{2} = 339V$

A sinusoidal a.c. supply has a peak voltage of 25V. It is connected across a 20Ω resistor.

Calculate the following:

- The r.m.s. voltage of the supply
- The r.m.s. current through the resistor
- The average power of the resistor

$V_{rms} = \dfrac{V_{peak}}{\sqrt{2}} = 17.7V$

$I_{rms} = \dfrac{V_{rms}}{R} = 0.884A$

$P = I^2 R = 15.6W$

Transformers

Imagine a power station is to supply power to a factory many miles away. In the diagram below the power supply represents the power station and the bulb represents the factory.

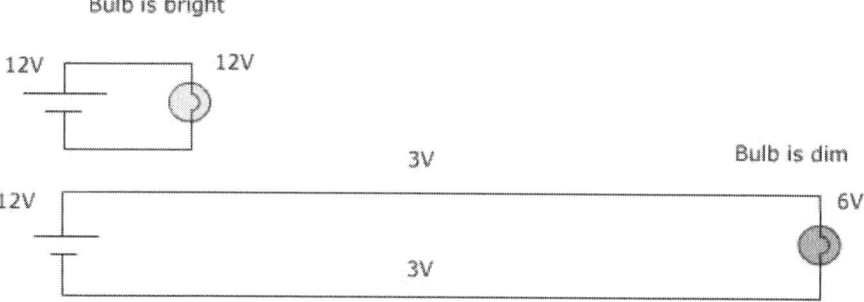

A certain amount of voltage, and therefore energy, is lost due to the resistance of the power lines. We could use very thick copper cables but these are expensive and long thick copper cables would not be strong enough to support their own weight.

Modern transmission cables are actually made of an aluminium core surrounded by steel cables. The aluminium is a reasonably good conductor and is light. The steel cables provide strength.

There is another solution.

A transformer can make an a.c. voltage bigger or smaller.

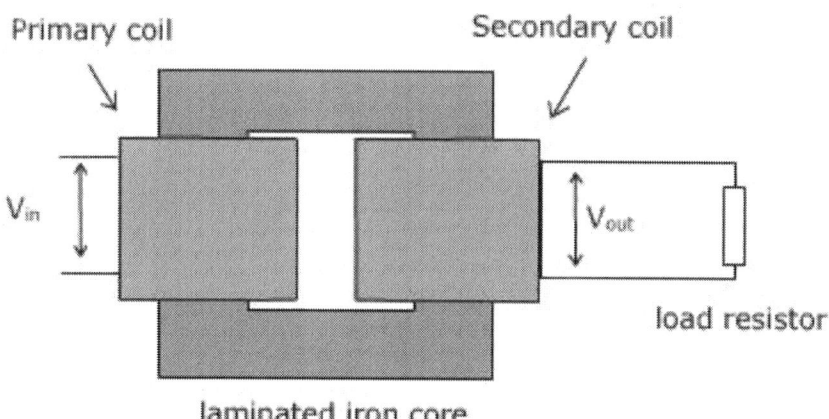

A transformer consists of two coils which share the same iron core.

- An alternating voltage is put cross the primary coil
- This produces an alternating magnetic field (flux) in the iron core
- An alternating e.m.f. is induced in the secondary coil

A step up transformer increases the voltage and decreases the current

A step down transformer decreases the voltage and increases the current

Whether the transformer steps up or steps down the voltage depends on the ratio of the number of turns on the coils.

The ratio of the turns is equal to the ratio of the voltages so:

$$\frac{V_p}{V_s} = \frac{N_p}{N_s}$$

For example if the secondary coil has 10 times more turns than the primary then the voltage would be stepped up by a factor of 10.

So why is voltage stepped up before being transmitted?

If the transformer were 100% efficient then the power input would equal the power output.

As $P = VI$ this means that if the voltage is stepped up then the current must be stepped down so if the voltage is 10 times bigger, the current is 10 times smaller.

$$\frac{V_p}{V_s} = \frac{N_p}{N_s} = \frac{I_s}{I_p}$$

The power loss in a supply cable of resistance R would be given by $P_{loss} = I^2 R$ so reducing the current would improve the efficiency of the transmission process significantly.

The voltage from the power station is stepped up (typically to 400,000V on the national Grid) before it is transmitted. It is stepped down to 240V before it is supplied to houses for safety reasons.

There is a fascinating bit of history concerning the rivalry between Thomas Edison and Nikola Tesla in the bid to electrify New York's street lighting. Edison's system used d.c. which, while being safer than high voltage a.c., had the major disadvantage that it could not be stepped up and so was much less efficient. Obviously Tesla eventually won the argument but not after a lot of dirty tricks on Edison's part which included public demonstrations of how dangerous a.c. was involving electrocuting animals.

Transformers are very efficient, typically around 97%, because they have no moving parts. There is some power loss due to heating of the coils and, as well as inducing an e.m.f. in the secondary coil, e.m.f.s will be induced in the iron core which is itself a good electrical conductor. These unwanted e.m.f.s will produce **eddy currents** in the core. These are swirling loops of current which produce heat and so reduce efficiency.

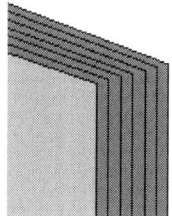

This is why transformer cores are laminated i.e. made up of thin layers of iron with an insulating resin between them. This means that large eddy currents cannot form.

A transformer consists of a primary coil with 50 turns and a secondary coil with 700 turns on an iron core. An alternating p.d. of 20V is put across the primary coil and a load resistance of 30Ω is connected across the secondary coil.

Calculate the following:

- The p.d. induced across the secondary coil
- The current drawn from the secondary coil
- The current in the primary coil
- The power input (assuming 100% efficiency)

This is a step up transformer with a turns ratio of 50:700 = 1:14

The output p.d. will therefore be 14 x 20 = 280V

The current drawn from the secondary $I = \dfrac{V}{R} = \dfrac{280}{30} = 9.33A$

The current from the primary will be 14 x 9.33 = 131A

Power input = power output = 131 x 20 = 2,620W

Questions

Q1

A magnetic field of strength 30mT flows through a circular coil with diameter 5cm. Calculate the flux linkage in the coil.

Q2

An iron core has a square cross section of sides 2cm. The total internal circumference is 30cm. It is made from iron which has a relative permeability of 300.

A coil with 100 turns is wrapped onto the core and carries a current of 0.6A.

Calculate the following:

- The permeance of the core
- The flux produced in the core by the current carrying coil
- The field strength inside the core

Q3

A long straight wire carrying a current of 2.5A lies at right angles to a magnetic field of strength 20mT

Calculate the force per metre acting on the wire

Q4

A 20 turn square coil with sides 5cm is placed in a uniform magnetic field of strength 50mT. Calculate the maximum torque that would act on the coil when a current of 0.9A flows through it.

What torque would act on the coil if the plane of the coil was at an angle of 30° to the magnetic field?

Q5

A proton travelling at 2×10^5 m/s passes through a uniform magnetic field of strength 0.3T with its velocity perpendicular to the field. (Mass of proton = 1.67×10^{-27} kg)

Calculate the force which would act on the proton.

Calculate the radius of the subsequent path of the proton

Q6

An electron is accelerated horizontally by a p.d. of 300V. The electron enters a vertical electric field of strength 10,000V/m and is deflected downwards. Calculate the strength of the magnetic field required to produce an upwards magnetic force to balance the electric force. (Ignore relativistic effects)

(Mass of electron = 9.1×10^{-31} kg)

Q7

A straight conductor of length 13cm moves at right angles to a magnetic field of strength 50mT at a velocity of 20m/s. calculate the e.m.f. induced across the ends of the conductor.

Q8

A bar magnet is dropped into a horizontal coil whose ends are connected to an oscilloscope.

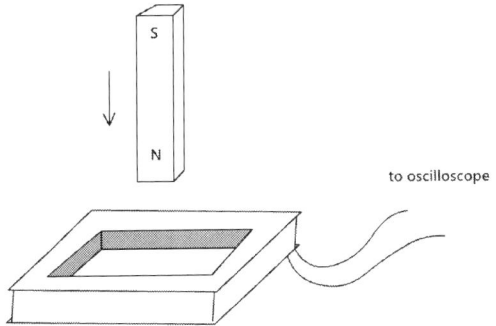

Sketch and explain the shape of the trace you would expect to see on the oscilloscope as the magnet falls through the coil.

Q9

Although copper and aluminium are not ferrous materials they can be separated using strong sweeping magnetic fields. Describe how this works.

Q10

A generator consists of a 200 turn rectangular coil with sides 5cm by 7cm which rotates at 300rpm inside a uniform magnetic field of strength 200mT.

Calculate the following:

- The angular velocity of the coil.
- The maximum flux linkage in the coil
- The peak e.m.f. induced in the coil
- The r.m.s. voltage produced by the generator

Q11

A signal generator produces the sawtooth waveform shown below with amplitude of 10V.

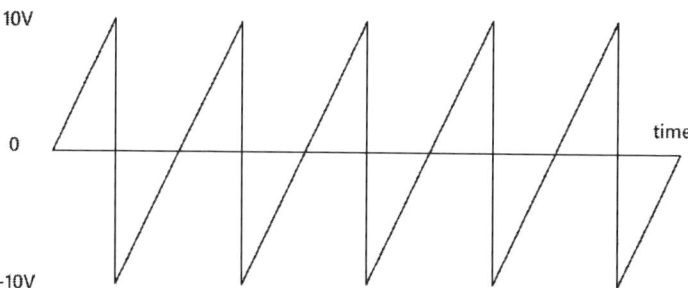

What would be the r.m.s. value of this waveform?

Q12

A step down transformer for a laptop is required to supply 20W of power at 12V from the 240V mains supply. Its primary coil has 500 turns.

Calculate the following:

- The current drawn from the mains supply
- The number of turns on the secondary coil
- The current drawn by the laptop from the transformer

Answers

Q1 $\Phi = BA$ $= 30 \times 10^{-3} \times \pi(0.025)^2 = 5.89 \times 10^{-5}$ Wb

Q2 $\Lambda = \frac{\mu A}{L}$ $= 5.03 \times 10^{-7}$ Wb/A turn

Q3 $\frac{F}{L} = BI$ $= 50$ mN/m

Q4   Couple = B I A N $= 2.25 \times 10^{-3}$ Nm   x sin 30 $= 1.13 \times 10^{-3}$ Nm

Q5   F = Bqv $= 9.6 \times 10^{-15}$ N $r = \frac{mv}{Bq}$ $= 6.92$ mm

Q6 $F_E = Eq$ $= 1.6 \times 10^{-15}$ N $v = 10.3 \times 10^6$ m/s $B = \frac{F}{qv} = 971 \mu T$

Q7 $\varepsilon = \frac{\Delta \varphi}{t} = \frac{BA}{t} = 0.13$ V

Q8

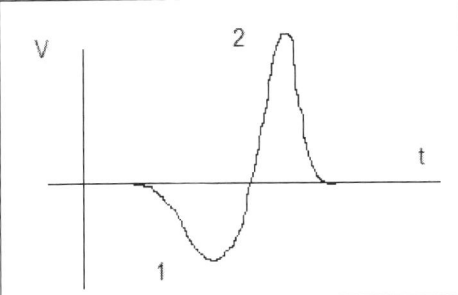

	Peak 1 As magnet enters emf opposes motion/flux cutting
	Peak 2 Narrower as magnet has accelerated due to gravity
	Bigger as rate of flux cutting is bigger
	In opposite direction to peak 1 as emf is opposing the magnet leaving the coil

Q9   A sweeping magnetic field induces emfs in conductors such as copper. This creates eddy currents. The eddy currents produce their own fields which oppose the change in flux. The conductors are therefore dragged off the conveyer belt.

Q10   ω = 31.4 rad/s   Nφ = 0.14 Wb turns $\varepsilon_{max} = 4.40$ V $V_{rms} = 3.11$ V

Q11   5V   If all the voltage was positive the average would be 5V

Q12 $I = \frac{P}{V}$ = 83mA   ratio = 20:1 N_s = 25 I_s = 1.6A

Printed in Great Britain
by Amazon